Authors and Reviewers

Tricia Salerno
Jenny Kempe
Bill Jackson
Allison Coates

Workbook

DIMENSIONS MATH 6A

 Singapore Math Inc.

Published by Singapore Math Inc.

19535 SW 129th Avenue
Tualatin, OR 97062
www.singaporemath.com

Dimensions Math® Workbook 6A
ISBN 978-1-947226-42-5

First published 2018
Reprinted 2019 (twice), 2020 (twice)

Printed in China

Contents

Chapter 3
Decimals

Chapter 4
Negative Numbers

Chapter 5
Ratios

Chapter 6
Rate

Chapter 7
Percent

Blank

1 Whole Numbers

1.1A Expressions and Equations

Basics

1. Identify the following as either an expression or an equation.

   (a) $27 + 27$ _____

   (b) $27 = 27$ _____

   (c) 27 _____

   (d) $27 \div 1$ _____

   (e) $1 = 27 \times 1 \div 27$ _____

2. Fill in the blanks.

   (a) The result of an addition expression is called the _____.

   (b) The result of a subtraction expression is called the _____.

   (c) The _____ is the result of a multiplication expression.

   (d) The _____ is the result of a division expression.

3. Match the expression to the statement.

   (a) 75 more than the quotient of 12 divided by 2   **1.** $75 - 5 \times 12$

   (b) 10 less than the product of 75 and 2   **2.** $75 \times 2 - 10$

   (c) 75 decreased by the product of 5 and 12   **3.** $12 \div 2 + 75$

4. Write an expression that matches each problem situation.
   Then evaluate each expression.

   (a) Michael bought 3 dozen eggs, but found that 5 eggs were broken.

   _____

   (b) Elena put a flower arrangement on each of 8 tables, and
   2 arrangements at the doorway.

   _____

   (c) Yara placed 2 bowls of chips on each of the 8 tables, and one bowl
   at the buffet table, the dessert table, and the drinks table.

   _____

5. Someone drew the following bar models to represent this statement: the product of 24 and 3 decreased by 5.
Write a two-step word problem which could accompany these bar models.

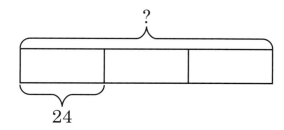

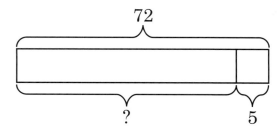

6. Write a word problem to accompany each expression, and then evaluate the expression.

(a) $15 \times 12 \div 3 + 4$

(b) $15 \times 12 \div (3 + 6)$

1 Whole Numbers

1.1B Exponents

Basics

7. Identify the base and the exponent by filling in the blanks.

   _____ $\longrightarrow 2^3 \longleftarrow$ _____

8. Fill in the blanks, and continue the pattern for 2^5.

   $2^2 = 2 \times 2 =$ _____

   $2^3 =$ _____ $\times$ _____ $\times$ _____ $= 8$

   $2^{—} =$ _____ $\times$ _____ $\times$ _____ $\times$ _____ $= 16$

   _____$^{—} =$ _____ $\times$ _____ $\times$ _____ $\times$ _____ $\times$ _____ $=$ _____

9. Evaluate the following:

   $1^4 =$ _____ $\times$ _____ $\times$ _____ $\times$ _____ $=$ _____

   $10^3 =$ _____ $\times$ _____ $\times$ _____ $=$ _____

   $100^2 =$ _____ $\times$ _____ $=$ _____

10. Write each expression using exponents and then find the value.

(a) 9×9 _____

(b) $8 \times 8 \times 8$ _____

(c) $3 \times 3 \times 3 \times 3 \times 3$ _____

(d) $3 \times 3 \times 3 \times 4 \times 4$ _____

(e) $5 \times 5 \times 2 \times 2 \times 2 \times 2$ _____

11. Evaluate the following expressions.

(a) 4^3 _____

(b) $4^2 \times 2^3$ _____

(c) $5^2 \times 4^2$ _____

(d) $3^3 \times 2^3$ _____

12. Leo evaluated 6^3 as 18. Explain his error and evaluate 6^3.

13. John squared a whole number, then multiplied the result by 10. His result was 810. What was the original number?

14. Students are attempting to show why we call a number multiplied by itself a "square number."
They find the following pattern using square tiles:

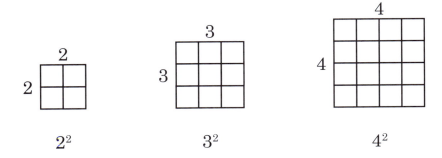

What would the model look like for 6^2?

15. Fill in the blanks to make each of the following equations true.

(a) $49 = 7\,\underline{}$

(b) $3\,\underline{} = 243$

(c) $3 \times 3^2 = 3\,\underline{}$

Basics

16. Circle the correct equation. Explain your thinking.

(a) $2 + 3 \times 5 = 17$

(b) $2 + 3 \times 5 = 25$

17. Solve the following equations.

(a) $9 \times 2 - 4 =$ _____

(b) $20 - 35 \div 7 =$ _____

(c) $1 \div 1 \times 1 + 1 - 1 =$ _____

18. (a) A sixth grade class was purchasing the prizes for the school carnival. They told Shop A that they needed 14 individual prizes and 12 packages containing 36 prizes each. Write an expression to represent the class order.

(b) The class is thinking about changing their order to 12 individual prizes and 14 packages containing 36 prizes each. Write an expression to show what their new order would look like.

(c) How many more prizes is the second order than the first?

19. For each phrase, write whether it comes first, second, or third according to the order of operations.

(a) Working from left to right, do all addition and subtraction.

_____

(b) Evaluate the exponents.

_____

(c) Working from left to right, do all multiplication and division.

_____

20. Evaluate the following expressions.

(a) $4^2 - 7 \times 2 + 8$ _____

(b) $50 + 3^4 \times 6 - 2^3$ _____

(c) $5^3 - 8 \times 7$ _____

(d) $23 + 5^2 \div 5 \times 32 - 7$ _____

(e) $4 + 3^2 - 3 \times 2$ _____

21. Holly baked 7 dozen chocolate chip cookies and 3 dozen oatmeal cookies for a fund-raiser. Before she delivered the cookies, her family ate 4 oatmeal cookies and 2 chocolate chip cookies.

(a) Write an equation to find the total number of cookies delivered. Solve.

_____

(b) A student found a different answer to **(a)** by not following the Order of Operations convention. What different answers could be obtained without following the convention? Show the work for any you find.

22. Solve the riddles below.

(a) If you add 8 to this number and then halve the result, you will get 10. _____

(b) If you multiply this number by 7 and then add 5 to the result, you will get 54. _____

(c) If you subtract 11 from this number and square the result, you will get 81. _____

(d) If you square this number and then raise the result to the third power, you will get 64. _____

1 Whole Numbers

1.1D Order of Operations with Parentheses

Basics

23. Write an expression with parentheses, when needed, for each statement.

(a) 48 divided by the sum of 3 and 5.

_____

(b) The quotient of 9 fives and 5 threes.

.

_____

(c) 2 times 3 times the difference of 15 and 9.

_____

24. Evaluate and match each expression to the correct value.

$2^2 + 2 \times (3 + 3^2) \div 3$ 24

$(2^2 + 2) \times (3 + 3^2) \div 3$ 13

$2^2 + (2 \times 3) + 3^2 \div 3$ 12

25. Evaluate the following expressions.

(a) $3 \times (2 + 6)$ _____

(b) $7^2 - 3 \times (8 - 5)$ _____

(c) $((16 \div 4)^2 + (2 \times 3^2)) \div 2$ _____

26. Insert parentheses to make the equations correct.

(a) $6 \times 3 + 3^3 = 180$ _____

(b) $1 = 1 + 39 \div 16 - 6 \div 2^2$ _____

(c) $5 + 7 \times 8 - 6 \times 2 = 48$ _____

27. Read the stories. Write expressions to match all of the actions in each story. Then evaluate the expressions.

    (a) Betty bought a book for $25, a pen for $4, and a scarf for $16. She paid with a fifty-dollar bill. How much change did Betty get?

    (b) For a concert, chairs were arranged in 7 rows with 15 chairs in each row. Then, people who arrived late took 11 chairs from another room and placed them in the back of the concert hall. How many chairs were in the concert hall altogether?

28. Cora and Alyssa solved the following expression independently and got different solutions:

    $6 + 3 \times 6 \div 2 + 4$

    Cora says that the solution is 19. Alyssa says that the solution is 31. Using the Order of Operations convention, which girl is correct? What was the other girl thinking?

Study the patterns below to answer questions 29 and 30.

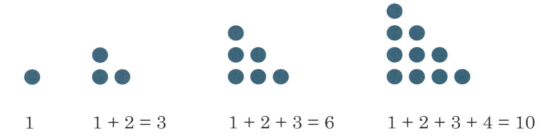

1   1 + 2 = 3   1 + 2 + 3 = 6   1 + 2 + 3 + 4 = 10

29. What is the number of dots in each of the next two patterns of dots?

30. Look for a pattern to find the sum of 1 + 2 + 3 ... 97 + 98 + 99 + 100.
Hint: You may choose to research Carl Friedrich Gauss.

_____

1 Whole Numbers

1.2A Factors

Basics

1. Find all of the factors of each of the following numbers.

   (a) 14 _____

   (b) 32 _____

   (c) 96 _____

   (d) 105 _____

2. What is the greatest common factor of the following numbers?

   (a) 36 and 45 _____

   (b) 28, 70 _____

   (c) 18, 36, 72 _____

3. **(a)** _____ and _____ are common factors of 30, 70, 114.

   (b) _____ and _____ are common factors of 210, 385, 462.

4. Taylor is making identical flower arrangements for a party. She has 21 yellow flowers, 35 pink flowers, and 42 white flowers. Each arrangement must have the same number of yellow, pink, and white flowers, and she must have some of each color in each arrangement. What is the greatest number of arrangements she can make using every flower? Describe the arrangements.

5. A tailor has one piece of fabric with a length of 60 m, and another piece of fabric with a length of 36 m. He wants to cut strips of equal length from the two pieces of fabric. How many meters long should each strip of fabric be if he wants the strips to be as long as possible?

6. Fang baked 32 lemon cookies, 40 peanut butter cookies, and 60 sugar cookies to put into bags. Each bag must have the same number of each type of cookie. What is the fewest number of bags Fang will need?

7. Multiply the numbers 11, 12, 13, ... 18 by 11. What patterns do you notice in the products?

8. Find the greatest common factor of 36, 60, 96, and 312.

Basics

9. List the first 5 multiples of each of the following numbers.

(a) 7 _____

(b) 12 _____

(c) 24 _____

10. List the first 10 multiples of 3 and 4. Circle the least common multiple.

11. Find the least common multiple of the following pairs of numbers.

(a) 12 and 18 _____

(b) 18 and 20 _____

12. Find the least common multiple of 4, 3, and 5. _____

13. Jett believes he can find the least common multiple of two numbers by simply multiplying the numbers together. Give an example of when he's right, and another example of when he's wrong.

14. The Chess club and the Robotics team both meet in the school gym. The Chess club meets every six days. The Robotics team meets every eight days. They are sharing the gym today. In how many days will they share the gym again?

15. A radio station is promoting a concert. Every 8th caller wins two tickets. Every 12th caller wins a backstage pass. Which caller will be the first to win both?

16. Mrs. Sanchez had red, pink, and blue pieces of ribbon. Each piece of ribbon was the same length. She cut the red ribbon into 2-in lengths, the pink ribbon into 7-in lengths, and the blue ribbon into 12-in lengths. What was the shortest possible length of the ribbons at the start if there was no remaining ribbon of any color?

17. Arthur, Brenda, and Carlos are wrapping gifts. Arthur wraps a gift every 3 minutes, Brenda wraps a gift every 4 minutes, and Carlos wraps a gift every 6 minutes. If they start wrapping a gift at the same time at 9:00 a.m., at what time will they start wrapping a gift at the same time again?

18. Alyssa has three sets of model trains. All three sets use the same train station, but use different train tracks. One of the trains takes 3 minutes to complete the track distance and return to the station. Another train takes 4 minutes to do the same. The third train takes 5 minutes to do the same. If all three trains leave the train station at the same time, how many minutes will it take until they meet at the station?

19. What number is 13 less than the least common multiple of 9, 12, and 15?

Basics

1. Write below each equation which property the equation demonstrates.

   (a) $a \times (b + c) = (a \times b) + (a \times c)$

   _____

   (b) $(a \times b) \times c = a \times (b \times c)$

   _____

   (c) $a \times 1 = a$

   _____

   (d) $a \times b = b \times a$

   _____

   (e) $a \times 0 = 0$

   _____

2. Evaluate the following multiplication expressions. Use the properties of multiplication to make it easier for you to find the answer.

    (a) $5 \times 7 \times 8 \times 2$ _____

    (b) $8 \times 9 \times 5$ _____

    (c) $(7 \times 5) \times (9 \times 2)$ _____

    (d) $100 \times (5 \times 12) \times (4 \times 7)$ _____

3. Fill in the blanks in the equations.

    (a) $7 \times 15 = 7 \times (10 + 5) =$ _____ $\times 10 +$ _____ $\times 5$

    (b) $7 \times 9 = 5 \times$ _____ $+ 2 \times$ _____

    (c) What property is used in the above equations?

4. The rectangles below have the same width but different lengths. Write an expression to show the total combined area of the two rectangles. Evaluate the expression.

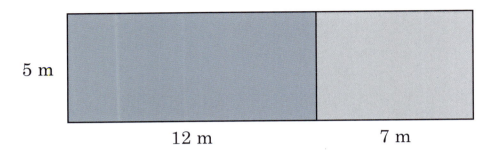

5 m

12 m   7 m

5. A muffin costs $2, a cake costs ten times as much as the muffin, and a bagel costs $0.50 less than the muffin.

   (a) Write an expression to show the total cost of 40 muffins, 3 cakes, and 5 bagels.

   (b) Jenna has $145. Does she have enough money to pay for the muffins, cakes, and bagels?

Basics

6. Use the distributive property to multiply mentally.
   Write expressions to show your thinking.

   (a) $24 \times 6 =$ _____

   (b) $65 \times 8 =$ _____

7. Use the distributive property to solve the following expressions.

   (a) 68×7 _____

   (b) 29×8 _____

8. Use the distributive and associative properties to break apart and
   rearrange the factors to solve the following expression.

   $6{,}000 \times 21$ _____

9. Use the distributive property to show how you could multiply the following mentally.

   (a) 38×6 _____

   (b) 36×7 _____

   (c) 8×126 _____

   (d) 7×49 _____

10. Show how you can apply properties of multiplication and division to solve these mentally.

    (a) $2,000 \times 24$ _____

    (b) 83×7 _____

Six students found the product of 26 and 8 mentally using different strategies. Their methods are represented by the strategies below.

Abigail

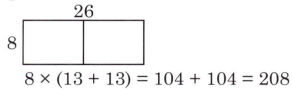

$8 \times (13 + 13) = 104 + 104 = 208$

Bron

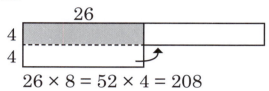

$26 \times 8 = 52 \times 4 = 208$

Chapa

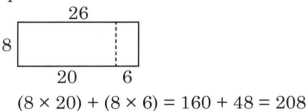

$(8 \times 20) + (8 \times 6) = 160 + 48 = 208$

Dexter

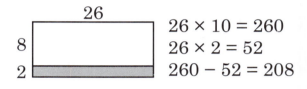

$26 \times 10 = 260$
$26 \times 2 = 52$
$260 - 52 = 208$

Ella

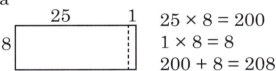

$25 \times 8 = 200$
$1 \times 8 = 8$
$200 + 8 = 208$

Franco

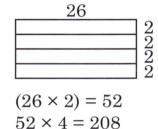

$(26 \times 2) = 52$
$52 \times 4 = 208$

11. Solve 16×7 using at least three different strategies.

Write your answers on the next page. You may choose to use some of the strategies shown on this page or use your own. Show your work.

(a)

(b)

(c)

(d)

(e)

(f)

Basics

1. Fill in the blanks using the following words.

   quotient   *dividend*   *divisor*

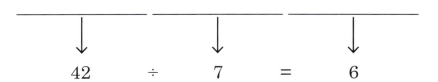

   42   ÷   7   =   6

2. Draw a model and an equation to match:

   (a) ***Sharing division problem.*** 18 yards of fabric is used equally to make 3 curtains. How many yards of fabric did it take to make each curtain?

   (b) ***Grouping division problem.*** Dan has 18 yards of fabric. How many curtains can he make if it takes 3 yd of fabric to make each curtain?

3. What happens to the quotient when you divide both the dividend and the divisor by the same non-zero number? Give two examples to support your answer.

_____

_____

4. Fill in the blanks.

(a) $42 \div 14 = (42 \div 7) \div (14 \div \underline{\hspace{1cm}}) = \underline{\hspace{1cm}}$

(b) $96 \div 16 = (96 \div 8) \div (16 \div \underline{\hspace{1cm}}) = \underline{\hspace{1cm}}$

(c) $15 \div 3 = (15 \times 2) \div (3 \times \underline{\hspace{1cm}}) = \underline{\hspace{1cm}}$

(d) $24 \div 6 = (24 \times \underline{\hspace{1cm}}) \div (6 \times 4) = \underline{\hspace{1cm}}$

5. Janet has 56 cubes and wants to put 8 cubes per bag in as many bags as she can. How many bags can she use? Draw a model and solve.

6. A rectangle with a length of 14 in is divided into two smaller rectangles, A and B, with areas of 72 in² and 40 in², respectively. Find the width of the rectangle.

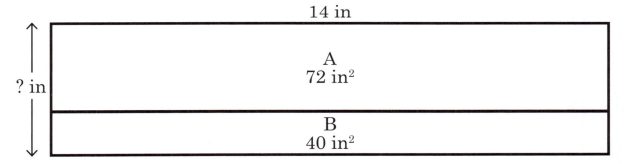

14 in

A
72 in²

B
40 in²

? in

7. Explain the Identity and Zero Properties of Division. Write three equations involving 100 to demonstrate both of these properties.

Basics

8. Fill in the blanks.

(a) Luis has 45 bagels. He has to put 3 bagels into each bag. How many bags will Luis use for all of the bagels?

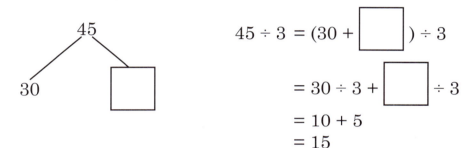

$$45 \div 3 = (30 + \boxed{}) \div 3$$
$$= 30 \div 3 + \boxed{} \div 3$$
$$= 10 + 5$$
$$= 15$$

Luis will use 15 bags.

(b) 50 trash cans hold a total of 3,550 lb of trash. How many lb of trash does each trash can hold?

$$3,550 \div 50 = (3,500 \div \boxed{}) + (50 \div 50)$$
$$= 70 + \boxed{}$$
$$= 71$$

Each trash can holds 71 lb of trash.

9. Mentally divide the following. Show your thinking.

   (a) $96 \div 8$

   (b) $102 \div 3$

   (c) $216 \div 3$

   (d) $285 \div 5$

   (e) $75,000 \div 250$

10. True or false? Explain your thinking.

    (a) $48,000 \div 16,000 = 24 \div 8$

    (b) $6,300 \div 2,100 = 21 \div 7$

11. $225 \div 45 = (225 \div \underline{\hspace{1cm}}) \div (45 \div \underline{\hspace{1cm}})$

    Think of 2 different whole numbers that could complete the equation.

12. Solve mentally.

    (a) 96 ÷ 6   _____

    (b) 204 ÷ 3   _____

    (c) 105 ÷ 7   _____

    (d) 4,900 ÷ 70   _____

    (e) 165,000 ÷ 15   _____

13. Mariam and Riya are both using mental math to evaluate 636 ÷ 6. Mariam says, "600 ÷ 6 = 100 and 36 ÷ 6 = 6, so the quotient is 106." Riya says, "636 ÷ 3 = 212 and 212 ÷ 2 = 106."

    Why do both girls' methods result in 106?

14. Tim is using his computer class to write his first app. He needs to figure out if his resolution is correct. Tim mentally computes that $1,920 \div 1,080$ is the same as $16 \div 9$. Is he correct? Explain how you know.

15. Calculate each of the following mentally.

(a) $2 \times (102 \div 3) - 5$ _____

(b) $5 \times (45,000 \div 5,000) \times 2$ _____

(c) $9^2 \div 3^2 \times 1,000$ _____

2 Fractions

2.1A Multiplication of a Proper Fraction by a Whole Number

Draw a bar model.
Solve.

1. $3 \times \frac{1}{5}$ _____

2. $\frac{3}{4} \times 2$ _____

3. Mary baked 3 batches of cookies. Each batch used $\frac{3}{4}$ cup raisins. How many cups of raisins did she use for all three batches? Draw a bar model. Solve.

4. Hector used $\frac{2}{5}$ meter of wood to make 1 shelf. How many meters of wood does Hector need to make 6 shelves? Draw a bar model. Solve.

5. Over the summer, Julia walked her favorite trail 48 times. The trail is $\frac{2}{3}$ mile long. How many miles did she walk on this trail over the summer? Simplify before multiplying. Solve.

6. Caleb has 345 friends on social media. $\frac{2}{3}$ of them are males. How many of these friends are females?

7. A car lot has 522 cars to sell. $\frac{5}{6}$ of the cars are new. How many are used?

8. There are 240 students in a school. $\frac{5}{8}$ of the students are girls. $\frac{1}{2}$ of the boys ride the bus to school. How many boys ride the bus to school?

9. Jasper had $424 altogether. He used $\frac{5}{8}$ of his money to buy a gift. He spent $\frac{1}{3}$ of the remainder on a book. How much more did Jasper spend on the gift than on the book?

Basics

10. Solve.

(a) $\frac{1}{4} \times \frac{1}{3}$ _____

(b) $\frac{1}{5} \times \frac{2}{3}$ _____

(c) $\frac{4}{5} \times \frac{5}{8}$ _____

(d) $\frac{6}{7} \times \frac{2}{3}$ _____

11. At a car dealership, $\frac{3}{5}$ of the vehicles for sale are trucks. 1 half of these trucks are white. What fraction of all of the vehicles for sale are white trucks?

12. Lauren has some jewelry pieces. $\frac{4}{7}$ of Lauren's jewelry pieces are necklaces. $\frac{1}{4}$ of the necklaces are made of beads. $\frac{1}{7}$ of the bead necklaces are made of green beads. What fraction of her jewelry pieces are necklaces made of green beads?

13. $\frac{3}{4}$ of the length of a $\frac{8}{9}$-m rod is painted red. $\frac{5}{8}$ of the rest of the rod is painted blue and the remaining section is not painted. What length of the rod is painted red and what length is not painted?

14. $\frac{1}{2}$ of a rectangular garden is used to plant flowers. $\frac{1}{4}$ of the part of the garden used to plant flowers is used to plant roses. $\frac{1}{3}$ of the part of the garden used to plant roses is used to plant red roses. What fraction of the entire garden is not used to plant red roses?

Basics

Evaluate the following expressions.

15. (a) $1\frac{2}{5} \times 3$ _____

(b) $3\frac{7}{8} \times 4$ _____

(c) $6 \times 1\frac{5}{12}$ _____

(d) $15 \times 2\frac{1}{10}$ _____

16. Wyatt has 4 packages to mail. Each package weighs $2\frac{2}{3}$ kg. What is the combined weight of the packages Wyatt wants to mail?

17. Linda bought 12 packages of seeds. Each package holds $2\frac{3}{8}$ oz of seeds. How many ounces of seeds did Linda buy?

18. A cabbage weighs $1\frac{1}{4}$ lb. A melon weighs three times as much as the cabbage. A pumpkin weighs twice as much as the melon. What is the weight of the pumpkin?

19. Josef collects sports cards. He has 45 football cards, and $\frac{5}{3}$ as many baseball cards as football cards. He has $\frac{3}{4}$ as many hockey cards as football and baseball cards combined. How many cards are in Josef's collection?

20. A school library has 1,200 fiction books. It has $1\frac{3}{4}$ as many nonfiction as fiction books. Of the nonfiction books, $\frac{1}{3}$ are history books. How many history books are in the school library?

Basics

21. Solve.

(a) $4\frac{5}{6} \times \frac{3}{4}$ _____

(b) $1\frac{7}{8} \times \frac{3}{4}$ _____

(c) $2\frac{5}{6} \times \frac{1}{3}$ _____

(d) $\frac{3}{4} \times \frac{16}{9}$ _____

22. Solve and give each answer in the simplest form.

(a) $\frac{18}{5} \times \frac{35}{9}$

(b) $1\frac{1}{3} \times \frac{15}{7}$

(c) $8\frac{1}{9} \times 1\frac{4}{5}$

23. Mr. King bought $3\frac{2}{3}$ tons of gravel for his yard. He used $\frac{1}{4}$ of it in the backyard. He used the rest in the side yard. How much gravel did he use in the side yard?

24. A rectangular field measures $6\frac{5}{8}$ ft by $2\frac{2}{3}$ ft. What is the area of this field?

25. Kamala had $4\frac{4}{5}$ L of juice. After giving $\frac{1}{4}$ of it to her brother, she drank $\frac{1}{9}$ of what she still had. She then used half of the remaining juice for a party. What is the difference between what Kamala had left and what she gave to her brother?

2 Fractions

2.2A Division of a Whole Number by a Fraction

Basics

1. Solve.

   (a) $12 \div \frac{1}{2}$ _____

   (b) $12 \div \frac{3}{2}$ _____

   (c) $12 \div \frac{5}{2}$ _____

2. (a) How many $\frac{2}{3}$s are in 6?

   (b) Divide 18 by $\frac{4}{7}$.

   (c) What is $5 \div \frac{1}{5}$?

3. Emily rode her bike a total of 10 miles. She stopped every $\frac{5}{4}$ mile. Including her last stop, how many stops did she make?

4. A bead company packed 27 lb of gold beads into bags of $\frac{3}{5}$ pound each. How many bags were there?

5. A baker has 1 kg of flour. One cup of the flour weighs $\frac{3}{25}$ kg. How many cups of flour does the baker have?

6. Ms. Aquino wants to pour 4 L of liquid equally into smaller containers, which hold $\frac{4}{5}$ L each. How many of these containers will Ms. Aquino fill?

Basics

7. Which of the following expressions is depicted by the model?

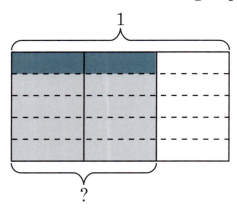

$\frac{2}{3} \div 3$ $\frac{2}{3} \div 5$ $\frac{2}{3} \div 15$

8. Solve.

(a) $\frac{4}{5} \div 20$ _____

(b) $\frac{5}{4} \div 3$ _____

(c) $\frac{1}{5} \div 5$ _____

9. Over the course of 10 years, a bush grew $\frac{5}{3}$ m. If the bush grew equal heights each year, how many meters did it grow each year?

10. Three friends shared $\frac{1}{2}$ of a pan of brownies equally. What fraction of a pan of brownies did each friend get?

11. Colton has $\frac{7}{8}$ lb of rice. He puts an equal amount of the rice into 7 bags. How many pounds of rice is in each bag?

12. A piece of ribbon is $\frac{7}{9}$ yd long. It is cut into 9 pieces. What is the length of each piece?

13. Emily spent $\frac{1}{5}$ of her savings on a washing machine. She then split $\frac{1}{2}$ of the remainder of her savings among her three children equally. What fraction of her savings did each child receive?

Basics

14. Solve.

(a) 15 thirds ÷ 5 thirds

_____

(b) 2 fourths ÷ 1 fourth

_____

(c) 3 fourths ÷ 1 fourth

_____

(d) 1 half ÷ 2 fourths

> Rename the half as fourths.

_____

15. Draw a model and solve.

(a) $\frac{7}{6} \div \frac{1}{6}$ _____

(b) $\frac{7}{6} \div \frac{2}{6}$ _____

(c) $\frac{7}{6} \div \frac{5}{6}$ _____

(d) $\frac{7}{6} \div \frac{6}{6}$ _____

16. Solve.

(a) $\frac{2}{6} \div \frac{1}{6}$ _____

(b) $\frac{10}{8} \div \frac{5}{8}$ _____

(c) $\frac{9}{10} \div \frac{2}{10}$ _____

(d) $\frac{11}{7} \div \frac{4}{7}$ _____

17. Solve.

(a) $\frac{3}{7} \div \frac{1}{2}$ _____

(b) $\frac{5}{8} \div \frac{7}{6}$ _____

(c) $1\frac{1}{8} \div \frac{1}{4}$ _____

(d) $1\frac{2}{3} \div 1\frac{3}{5}$ _____

18. Write a word problem, show a model, and solve for $\frac{3}{4} \div \frac{1}{2}$.

19. Pablo has a $2\frac{4}{5}$ yd of rope. He cuts the rope into pieces measuring $\frac{2}{5}$ yd each. How many pieces of rope does he end up with?

20. A turtle could crawl $\frac{3}{5}$ of a mile in $\frac{2}{3}$ of an hour. How far could the turtle crawl in one hour?

21. The area of a rectangular picture is $12\frac{1}{2}$ ft². The width of the picture is $2\frac{3}{4}$ ft. What is the length of the picture?

22. The area of a rectangular garden is $26\frac{1}{2}$ ft². The length of the garden is $12\frac{1}{2}$ ft. How much fencing would be needed to fence the entire garden?

Basics

Fill in the blanks.

1. **(a)** $\frac{35}{100}$ = _____ tenths _____ hundredths

    (b) $\frac{1}{1,000}$ = _____ tenths _____ hundredths _____ thousandth

    (c) $6\frac{2}{100}$ = _____ ones _____ tenths _____ hundredths

    (d) $\frac{87}{10}$ = _____ ones _____ tenths

    (e) $\frac{44}{100}$ = _____ ones _____ tenths _____ hundredths

2. Write the following as decimals.

    (a) $\frac{9}{10}$ _____   **(b)** $\frac{23}{10}$ _____

    (c) $\frac{45}{100}$ _____   **(d)** $\frac{52}{10}$ _____

    (e) $\frac{305}{100}$ _____   **(f)** $\frac{32}{1,000}$ _____

3. Write the following as decimals.

(a) $\frac{3}{4}$ _____

(b) $\frac{44}{50}$ _____

(c) $3\frac{1}{5}$ _____

(d) $\frac{9}{8}$ _____

(e) $\frac{15}{4}$ _____

(f) $4\frac{15}{20}$ _____

4. Express each decimal as a fraction or mixed number in simplest form.

(a) 6.17 _____

(b) 3.60 _____

(c) 0.24 _____

(d) 13.225 _____

(e) 2.45 _____

(f) 8.028 _____

5. Write the following as decimals.

    (a) 3 tenths 2 thousandths   _____

    (b) 45 tenths   _____

    (c) 2 tens 72 thousandths 3 tenths   _____

6. Arrange these numbers in increasing order.

    (a) 0.1 $\frac{1}{100}$ 0.001 $1\frac{1}{10}$

    _____ , _____ , _____ , _____

    (b) $16\frac{5}{25}$ 16.02   16.022 $16\frac{22}{100}$

    _____ , _____ , _____ , _____

7. Find three even five-digit numbers in which the digit in the thousandths place is the same as the digit in the ones place, the digit in the hundredths place is greater than the digit in the tenths place, and the sum of the digits is 16.

3 Decimals

3.1B Adding and Subtracting Decimals

Basics

8. Calculate the following.

   (a) $4.209 + 17.096$ _____

   (b) $81.4 - 3.629$ _____

   (c) $16.02 + 11 + 0.005$ _____

   (d) $10 - 0.901 - 3.7$ _____

9. Express the value of the following as a decimal.

   (a) $1.06 + \frac{5}{10}$ _____

   (b) $32.5 - 30\frac{1}{4}$ _____

   (c) $1\frac{2}{10} + 0.03 + \frac{4}{1,000}$ _____

   (d) $\frac{1}{4} - 0.025$ _____

10. Grace buys a scarf for $23.88, a blouse for $42.09, and a skirt for $33.77. She has a gift card worth $100. What will be the value of Grace's gift card after she buys these items?

11. Hannah has some scraps of ribbon measuring 3.52 meters, 17.3 meters and 27.009 meters. Does she have enough to do a project requiring 50 meters of ribbon? If yes, how much ribbon will she have left over? If no, how much more ribbon does she need?

12. One trail through the park is 6.45 miles long. Another trail is 2.08 miles longer than the first trail. What is the combined distance of both trails?

13. Fill in the blanks.

    (a) 3.074 =_____ones +_____tenths +_____hundredths +

        _____thousandths

    (b) 3.074 =_____thousandths

    (c) 3.074 =_____hundredths +_____thousandths

    (d) 3.074 =_____tenths +_____hundredths +_____thousandths

14. The difference between two numbers is 4.26. Their sum is 28.18. What are the two numbers?

15. A runner ran 3.247 km on Monday, 3.9 km on Tuesday, 4.12 km on Wednesday, 3.006 km on Thursday, and 0.9 km farther on Friday than on Monday. If the runner's goal for the five days was to run 20 km, by how much did he miss his goal?

16. A school spent \$2,296 on two identical computers and three identical printers. Each computer cost \$610.50 more than a printer. How much less did the school spend on printers than on computers?

17. Two books and three pens cost $41.25. Four books and five pens cost $79.75. How much does one pen cost?

3 Decimals

3.2A Decimal Number System

Basics

1. Solve.

   (a) 0.005×10 _____

   (b) 0.005×100 _____

   (c) $0.005 \times 1{,}000$ _____

   (d) $500 \times \frac{1}{10}$ _____

   (e) $500 \times \frac{1}{100}$ _____

   (f) $500 \times \frac{1}{1{,}000}$ _____

2. Solve.

   (a) $4 \text{ tens} \times 10$ _____

   (b) $4.2 \text{ tens} \times 100$ _____

   (c) $71.7 \div 10$ _____

   (d) $\frac{1}{10} \times 71.7$ _____
   (Hint: $71.7 \div 10$)

   (e) $9 \div \frac{1}{100}$ _____

   (f) 0.09×0.1 _____

3. Eli has saved $48.50.

(a) His older sister has saved 10 times as much. How much money has his older sister saved?

(b) Eli's brother borrowed $\frac{1}{10}$ of Eli's savings. How much money does Eli have left?

4. Koni spent \$78.50 on a guitar. She spent 100 times as much on a piano than on the guitar. She spent $\frac{1}{10}$ as much on a violin than she did on the piano. How much did she spend on all three instruments?

Basics

5. Without calculating, determine if the following multiplication expressions will give a product that is more or less than 205.63. Explain your answers.

(a) 205.63×0.79

(b) 205.63×1.01

(c) 205.63×0.998

(d) 205.63×1.3

6. Fill in each blank with the correct symbol: >, <, or =.

(a) 6.5 _____ 6.5 × 1

(b) 1.5 × 6.5 _____ 6.5

(c) 6.5 × 0.15 _____ 6.5

(d) 0.999 × 6.5 _____ 6.5

(e) 6.5 × 1.001 _____ 6.5

(f) 6.5 _____ 6.5 × 9.9

(g) $6.5 \times \frac{1}{1}$ _____ 6.5

(h) 6.5 _____ $\frac{9}{10} \times 6.5$

(i) $\frac{99}{10} \times 6.5$ _____ 6.5

Basics

7. Write each expression in vertical form, then multiply.

(a) 1.5×3 _____

(b) 17×3.3 _____

(c) 0.5×0.7 _____

(d) 0.34×0.23 _____

(e) 4.1×3.2 _____

8. Solve. Show your work.

   (a) Find the area of a square tablecloth with 42.5-inch sides.

   (b) Find the volume of a cube-shaped fish tank with sides measuring 2.7 meters.

   (c) Isaac is reupholstering a boat cushion. He needs 3.2 yards of nautical fabric which costs $18.27 per yard. How much money will he spend on fabric? Round the answer to the nearest cent.

9. A bean plant grown in the shade is 15.3 cm tall. A similar bean plant grown in the sun is 3.8 times as tall as the first bean plant. How many cm shorter is the shade-grown bean compared to the sun-grown bean?

10. Alyssa saved $5,280. She spent $\frac{3}{10}$ of it on furniture and $\frac{1}{5}$ of it on a computer. She then spent $\frac{1}{4}$ of the remainder on clothes. How much money did Alyssa have in the end?

11. Anna saved $45.65 one week. The next week, she saved twice as much as the first week. The third week, she saved $10 less than three times as much as the second week. How much money did Anna save altogether in the three weeks?

Basics

1. Without doing any calculations, determine which of the following division expressions will have a quotient that is less than the dividend. Explain your answers.

    (a) $20.48 \div 2$

    (b) $20.48 \div 0.2$

    (c) $20.48 \div 12$

    (d) $20.48 \div 1.2$

2. Fill in each blank with the correct symbol: >, <, or =.

(a) $6.5 \rule{2cm}{0.4pt} 6.5 \div 1$

(b) $6.78 \div 6.5 \rule{2cm}{0.4pt} 6.5$

(c) $6.5 \rule{2cm}{0.4pt} 6.5 \div 0.15$

(d) $6.5 \div 6.5 \rule{2cm}{0.4pt} 6.5$

(e) $6.5 \div 1.001 \rule{2cm}{0.4pt} 6.5$

(f) $6.5 \rule{2cm}{0.4pt} 6.5 \div 9.9$

(g) $6.5 \div \frac{1}{100} \rule{2cm}{0.4pt} 6.5$

(h) $6.5 \rule{2cm}{0.4pt} 6.5 \div \frac{98}{10}$

(i) $6.5 \div \frac{99}{100} \rule{2cm}{0.4pt} 6.5$

Basics

3. Replace the ? to make the expressions equivalent.

(a) $8 \div 4 = (8 \times 3) \div (4 \times ?)$ _____

(b) $7.2 \div 9 = (? \times 10) \div (9 \times ?)$ _____ _____

(c) $3.69 \div 0.6 = 36.9 \div ? = 6.15$ _____

(d) $30.9 \div 0.15 = ? \div 15 = 206$ _____

4. Solve.

(a) $72 \div 9$ _____

(b) $7.2 \div 9$ _____

(c) $0.72 \div 9$ _____

(d) $0.72 \div 0.9$ _____

(e) $72 \div 0.09$ _____

5. Solve. Round the quotients to the nearest tenth.

    (a) Divide 8 by 16 _____

    (b) $1.16 \div 0.8$ _____

    (c) $0.372 \div 0.03$ _____

    (d) Divide 8.866 by 0.13 _____

    (e) $\frac{156}{4.8}$ _____

    (f) Divide 0.393 by 0.15 _____

    (g) $\frac{16.8}{5.4}$ _____

6. It took 20 tons of gravel to improve a 2.5-mile section of a path in the park. At this rate, how much gravel will it take to improve each mile?

7. The area of a garden is 330.72 m². The length of the garden is 15.6 m. What is the width of the garden?

8. Farmer Chang grew 1,085.7 kg of kale. He packaged it in bags that each held 51.7 kg of kale. How many bags did he package?

9. I need to buy laundry detergent. I could buy the jumbo size or the medium size. The jumbo size costs $17.94 and holds 4.43 liters. The smaller size costs $12.98 and holds 2.95 liters. If I am looking for a bargain, which one should I buy?

10. A metal pipe measuring 2.75 m weighs 13.75 kg. This type of pipe costs $16.50 per kg. How much would 2 m of this type of pipe cost?

3 Decimals

3.4 Metric Measurements and Decimals

Basics

1. Solve.

(a) 15 km = _____ m

(b) 1.5 km = _____ m

(c) 0.15 km = _____ m

(d) 0.15 km = _____ cm

(e) 7 kg = _____ g

(f) 0.7 kg = _____ g

(g) 62 L = _____ mL

(h) 6.2 L = _____ mL

2. Express the following as a decimal.

(a) 5 mL = _____ L

(b) 5.2 mL = _____ L

(c) 13 g = _____ kg

(d) 13.8 g = _____ kg

(e) 0.8 m = _____ km

(f) 8.8 m = _____ km

(g) 1,060.2 cm = _____ m

(h) 1,200 mm = _____ m

3. Hazel had 8.3 m of fabric. She used 5.9 m to make curtains for her clubhouse. She needs 80 cm of fabric to make a cushion. How many cushions can Hazel make?

4. Box A weighs 8.5 kilograms. Box B weighs 7.35 kilograms. Box C weighs 980 grams.

(a) How much do all 3 boxes weigh in kilograms?

(b) What is the difference in weight between Box A and Box C in grams?

5. Ms. Fujimoto has three barrels of water. Each barrel holds 4.6 liters. Ms. Fujimoto must transfer the water equally from the barrels to 24 smaller, equally-sized containers. How many milliliters of water should she put in each small container?

6. Rea wants to mail five gifts. Four gifts are identical, and each weighs 1.8 kg. The total contents of the shipping box cannot exceed 10 kg. What is the maximum weight, in grams, that the fifth gift can weigh?

7. Two similar jars and two similar bottles can hold 6.3 liters of liquid. Five such jars and four such bottles can hold 14.6 liters of liquid. How much liquid can one bottle hold?

4 Negative Numbers

4.1 Positive and Negative Numbers

Basics

1. Express the following real world situations using positive and negative numbers. Give the reference point for each situation. The first one is done for you.

   (a) The position of a sunken ship 3,000 meters below sea level.

   Reference point:_____*sea level*_____   _____*−3,000 m*_____

   (b) A temperature of 85°F.

   Reference point:_____   _____

2. If −6 represents 6 m below sea level, then +20 represents

   _____.

3. Suppose a new calendar marks 2000 AD as 0 DM. What year in DM is 1996 AD? What year in DM is 2002 AD?

   (a) What number stands for the year 1996?   _____

   (b) What number stands for the year 2002?   _____

4. The average weight of an adult male Rufous hummingbird is 3.2 grams. Below is a table showing the weights of five male Rufous hummingbirds. Express the difference between each bird's weight and the average weight, using positive or negative numbers.

Bird A	Bird B	Bird C	Bird D	Bird E
3.25 g	3.22 g	3.19 g	3.195 g	3.201 g

Bird A _____   Bird B _____

Bird C _____   Bird D _____

Bird E _____

5. The average amount the five Marsh children have been able to save is $25. Below is a table showing each child's savings. Express the difference between each child's savings and the average savings using positive or negative numbers.

Ada	Briana	Cooper	Darryl	Ethan
$18.13	$27.08	$6.15	$40.37	$33.27

Ada _____   Briana _____

Cooper _____   Darryl _____

Ethan _____

6. Over the course of several weeks, Lauren hiked five trails in Joshua Tree National Park. Her average distance was 5.34 miles. Using the table below, express the difference between the distance of each hike and the average distance. Express the difference in positive or negative numbers.

49 Palms Oasis	Split Rock Loop	Lost Palms Oasis	Lost Horse Loop	Willow Hole
3 miles	2.5 miles	7.5 miles	6.5 miles	7.2 miles

49 Palms Oasis _____   Split Rock Loop _____

Lost Palms Oasis _____   Lost Horse Loop _____

Willow Hole _____

7. The Canton Tower in Guangzhou, China is 600 meters tall. Willis Tower in Chicago, Illinois, USA is 443 meters tall.

(a) Which building is taller, and by how much?

(b) A man standing on the top of the Willis Tower is fixing an antenna. Express the position of the man using the top of the Canton Tower as a reference point.

8. The four deepest subway stations in New York City are listed in the table below.

Station	Feet Below Street Level
191st Street	180 ft
190th Street	140 ft
Roosevelt Island	100 ft
Lexington Avenue at 63rd Street	100 ft

(a) A commuter is in the 190th Street Station. Express the position of the commuter using the Roosevelt Island Station as a reference point.

(b) Another commuter is in the 191st Street Station. Express the position of this commuter using 190th Street Station as a reference point.

(c) From which station to which reference point would be the greatest distance?

4 Negative Numbers

4.2A The Number Line

1. Draw a horizontal number line from −6 to 3, labeling integer increments.

   (a) Represent the numbers 0, −5, 2.5, −2$\frac{1}{2}$ on the number line.

   (b) Arrange the numbers in ascending order, from least to greatest.

   _____

   (c) Write two numbers that are less than −2$\frac{1}{2}$. _____

2. Use the vertical number line shown to do the following.

   (a) Represent the numbers −3.2, −2.4, 1$\frac{1}{2}$, 1$\frac{1}{5}$.

   (b) Arrange the numbers in ascending order, from least to greatest.

   _____

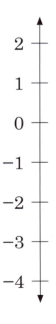

3. Place an inequality sign (> or <) in the blanks to make each statement true.

(a) 0 _____ −2

(b) −3.2 _____ −3

(c) 2 _____ $-2\frac{7}{8}$

4. Underline the correct answer for each of the following questions.

(a) Which is further from sea level: 2,675 m or −2,657 m?

(b) Which temperature is colder: −10 °F or 5 °F?

(c) Which represents a greater debt: −$1,350 or −$1,503?

5. The average score on a math quiz is 40 points. The table below shows the scores of 5 of the students who took the quiz.

Ani	Shanice	Nolan	Tyler	Mia
46	38	45	32	30

Express the difference between each student's score and the average score on the quiz, using positive or negative numbers.

Ani	Shanice	Nolan	Tyler	Mia

6. The table below shows the monthly rainfall, in inches, for nine months in Seattle, Washington.

Jan	Feb	Mar	Apr	May	Jun	Jul	Aug	Sep
$5\frac{1}{2}$	$4\frac{1}{4}$	$3\frac{11}{12}$	$3\frac{5}{12}$	$3\frac{3}{4}$	$2\frac{5}{12}$	$\frac{11}{12}$	$1\frac{1}{8}$	$1\frac{3}{4}$

The average monthly rainfall that year in Seattle was $3\frac{5}{12}$ in.

Express the difference between each month's rainfall, in inches, and the average monthly rainfall using positive or negative numbers. Express your answers in simplest form.

Jan	Feb	Mar	Apr	May	Jun	Jul	Aug	Sep

7. Arrange the following numbers in order from greatest to least.

   $-4\frac{1}{3}$, -4.25, 6, $4\frac{1}{2}$, 0.4, -0.4

   _____

8. Name one other fraction and one other decimal that would fall in the range of the numbers listed in question 7.

   _____

4 Negative Numbers

4.2B Absolute Value

Basics

9. Willis Tower is 443 meters above sea level. Assuming sea level is at 0 meters, express the height of Willis Tower using positive or negative numbers.

10. Use the number line below to answer the questions (a) through (e).

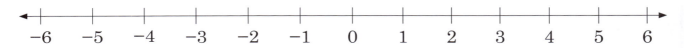

-6 -5 -4 -3 -2 -1 0 1 2 3 4 5 6

   (a) How many units is the distance from 0 to 4? _____

   (b) How many units is the distance from 0 to −4? _____

   (c) How many units is the distance from 0 to 6? _____

   (d) How many units is the distance from 0 to −6? _____

   (e) What is the total distance from −4 to 4? _____

11. Fill in the blanks with >, <, or =.

   (a) −135 _____ 22   **(b)** 36 _____ −63   **(c)** −57 _____ −42

   (d) −75.5 _____ −75.6   **(e)** $-4\frac{1}{2}$ _____ −4.5   **(f)** $-567\frac{1}{3}$ _____ $-765\frac{2}{3}$

12. Fill in the blanks with >, <, or =.

(a) $|-135|$ _____ $|22|$

(b) $|36|$ _____ $|-63|$

(c) $|-57|$ _____ $|-42|$

(d) $|-75.5|$ _____ $|-75.6|$

(e) $|-4\frac{1}{2}|$ _____ $|-4.5|$

(f) $|-567\frac{1}{3}|$ _____ $|-765\frac{2}{3}|$

13. Arrange the following numbers in order from least to greatest.

$|-72.3|$, $|7.2|$, $|-0.8|$, $|1|$, $|73.2|$

_____

14. Arrange the following numbers in order from greatest to least.

$|-12|$, $|12|$, $|8|$, $|21|$, $|-8|$, $|-11|$

_____

15. In a sixth grade math class, the teacher has created this grading rubric on tests:

Correct answers: 5 points
Incorrect answers: −2 points
Unanswered: 0 points

One of the teacher's tests consisted of 10 questions.

(a) What is the maximum score on the test? _____

(b) What is the minimum score on the test? _____

(c) Write a solution where a student scores 40 points on the test.

_____

(d) Write a solution where a student scores 36 points on the test.

_____

16. (a) What is the value of $-|-4|$? _____

(b) $|-3| + |-5| = $ _____

5 Ratios

5.1A Finding Ratio

Basics

1. Look at the shapes.

   (a) What is the ratio of hearts to stars? _____

   (b) What is the ratio of stars to clouds? _____

   (c) What is the ratio of stars to hearts? _____

   (d) What is the ratio of clouds to stars to hearts? _____

   (e) What is the ratio of hearts to the total number of shapes? _____

   (f) Write a question and a ratio that matches the picture.

2. For every 3 pears a grocery store sells, they sell 8 oranges.

   (a) What is the ratio of pears to oranges? _____

   (b) What is the ratio of oranges to pears? _____

   (c) What is the ratio of pears to the total number of fruit?

   _____

3. Chapa is making fruit punch by mixing 3 liters of grape juice with 2 liters of sparkling water.

   (a) What is the ratio of grape juice to sparkling water?

   _____

   (b) What is the ratio of sparkling water to grape juice?

   _____

   (c) What is the ratio of fruit punch to sparkling water?

   _____

4. A teacher bought 15 books. 8 of the books were fiction and the rest were nonfiction.

   (a) The ratio of nonfiction to fiction books is _____ : _____.

   (b) The ratio of fiction books to the total number of books bought is

   _____ : _____.

5. A rectangle has a length of 3 yards and a width of 2 feet. What is the ratio of the length of the rectangle to its width in feet?

_____

6. Evan is 3 years 1 month old. His sister is 5 years old. Express the ratio of Evan's age to that of his sister using whole numbers.

_____

7. The table below shows the types and numbers of vehicles in a parking lot.

Types of Vehicles	Convertibles	Coupes	Minivans	Sedans	Trucks
Number	3	7	13	11	9

(a) What is the ratio of the number of minivans to the number of all the other vehicles combined?

_____

(b) What is the ratio of the number of convertibles to all the vehicles in the parking lot?

_____

8. John and Amy had the same amount of money at first. After John spent $8.75 and Amy spent $29.25, the ratio of John's money to Amy's money was 5 : 3. How much money did Amy have at first?

9. Adam's sister is 4 years younger than Adam. The current ratio of Adam's age to his sister's age is 5 : 3. What will the ratio of Adam's age to his sister's age be in 5 years time?

Basics

10. There are groups of triangles as shown below.

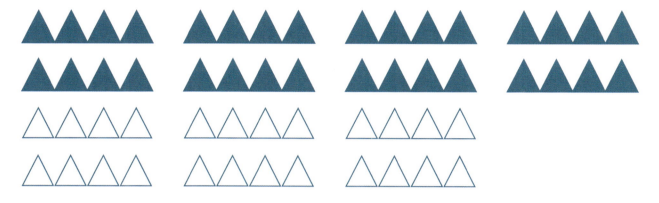

(a) What is the ratio of the number of colored triangles to the number of white triangles? Express the ratio in simplest form.

_____

(b) What is the ratio of white triangles to the number of colored triangles? Express the ratio in simplest form.

_____

11. Draw lines to match equivalent ratios.

(a)   2 : 1   **(e)**   2 : 3

(b)   6 : 9   **(f)**   100 : 30

(c)   7 : 5   **(g)**   8 : 4

(d)   10 : 3   **(h)**   35 : 25

12. Express each ratio in simplest form.

(a) $6 : 18$ _____

(b) $16 : 24$ _____

(c) $3 : 12$ _____

(d) $24 : 42 : 6$ _____

(e) $16 : 12$ _____

(f) $24 : 27$ _____

(g) $63 : 45$ _____

(h) $28 : 84 : 70$ _____

13. Find the missing term in the equivalent ratios.

(a) $3 : 5 =$ _____ $: 15$

(b) $20 : 32 = 5 :$ _____

(c) $4 :$ _____ $= 5 : 10$

(d) _____ $: 25 : 15 = 2 :$ _____ $: 3$

14. At a farm stand, there are 16 crates of tomatoes and 40 crates of potatoes.

    (a) What is the ratio of the number of crates of tomatoes to the number of crates of potatoes? Express the ratio in simplest form.

    _____

    (b) What is the ratio of the number of crates of potatoes to the total number of crates of potatoes and tomatoes? Express the ratio in simplest form.

    _____

15. Three friends were competing to see who would walk the furthest distance in a week.
Lincoln walked 20 miles. Amelia walked 12 miles less than Lincoln, and 20 miles less than Jett. Find the ratio of the distance Lincoln walked to the distance Amelia walked to the distance Jett walked.

16. Franco has 126 sports cards. Fifty-four of the cards are baseball cards, 30 are hockey cards, and the rest are football cards. What is the ratio of the football cards to the baseball cards to the hockey cards?

17. For singles matches, a tennis court is 78 feet long and 9 yards wide.

    (a) In feet, what is the ratio of the length of the court to its width?

    _____

    (b) In feet, what is the ratio of the width of court to its length?

    _____

    (c) In yards, what is the ratio of the length of the court to its width?

    _____

    (d) If you use yard as the unit to calculate the ratio of the width to its length, will the answer be different from the answer for (b)?

    78 ft

    _____

    9 yd

18. The ratio of the lengths of the sides of a triangle is 3 : 4 : 5. If the length of the shortest side of the triangle is 12 cm, what is the perimeter of the triangle?

19. The ratio of the lengths of two pieces of ribbon is 2 : 7. The shorter piece of ribbon is 12 m long. Find the total length of the two pieces of ribbon.

20. A sum of money is shared between Marco and Madeline in the ratio 2 : 3. If Madeline were to give $16 to Marco, they would have the same amount of money. How much money does Marco have?

21. A robotics club had a ratio of the number of girls to the number of boys of 5 : 3. After another 6 girls joined the club, the ratio became 2 : 1. How many girls were in the club at first?

5 Ratios

5.2 Ratios and Fractions

Basics

1. Express each ratio as a fraction in simplest form:

    (a) $1 : 4$ _____

    (b) $12 : 18$ _____

    (c) $14 : 5$ _____

    (d) $25 : 10$ _____

2. The ratio of the length of Ribbon A to the length of Ribbon B is $5 : 4$. Express the ratio of the length of Ribbon A as a fraction of the length of Ribbon B.

3. There are $\frac{3}{5}$ as many apples as pears. What is the ratio of the number of apples to the numbers of pears?

    _____

4. Alana and Jerry share stamps in the ratio of 7 : 9.

   (a) What fraction of the stamps did Alana get?

   (b) What fraction of the stamps did Jerry get?

   (c) If Jerry got 27 stamps, how many stamps did Alana get?

5. The ratio of the number of forks to the number of spoons in a drawer is 3 : 5.

    (a) What fraction of the eating utensils are forks?

    (b) What fraction of the eating utensils are spoons?

    (c) If there are 15 forks in the drawer, how many spoons are there in the drawer?

6. Paula and Kawai shared $312 in the ratio of 6 : 7. How much money did each person get?

7. A recipe calls for $1\frac{1}{2}$ cups water, $\frac{1}{4}$ cup butter, $\frac{1}{8}$ cup sugar, and 2 cups flour. What is the ratio of water to butter to sugar to flour?

8. Mr. Nakamura is $\frac{9}{5}$ as tall as his granddaughter. Mr. Nakamura is 180 cm tall.

    (a) Find the ratio of Mr. Nakamura's granddaughter's height to Mr. Nakamura's height.

    (b) Find the height of Mr. Nakamura's granddaughter.

9. Santiago and Amanda share 42 sports cards in the ratio 2 : 5. The ratio of the number of Amanda's hockey cards to the number of baseball cards is 2 : 3. How many more baseball cards than hockey cards does Amanda have?

10. The ratio of Eli's age to Jamal's age is 3 : 4. The ratio of Jamal's age to Avery's age is 2 : 3.

(a) Express Eli's age as a fraction of Avery's age. Give your answer in its simplest form.

(b) If Jamal is 28 years old, how many years older than Eli is Avery?

11. $\frac{2}{9}$ of the bikes in a bicycle shop are mountain bikes. The remaining bikes are road bikes and cruiser bikes in the ratio of 9 : 5. There are 15 more cruiser bikes than mountain bikes. How many road bikes are in the bicycle shop?

12. The ratio of the number of beads that Jenna and Koni had was 3 : 5. Jenna gave Koni $\frac{1}{6}$ of her beads. What is the fraction of Jenna's beads to Koni's beads now?

6 Rate

6.1 Average and Rate

1. The table shows the points Cooper got on 4 tests. In order to pass the course, he needs an average of 7 points.

Test 1	Test 2	Test 3	Test 4
9	8	6	9

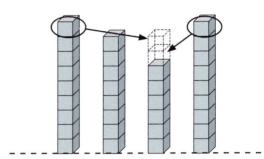

Did Cooper pass the course?

2. Find the mean of this set of numbers.

42, 78, 81, 23, 12

3. Find the average of each of the following set of measurements:

(a) 82 cm, 64 cm, 98 cm, 44 cm

(b) 3.25 L, 7 L, 2.5 L, 8.45 L, 1.75 L

4. Four dogs have weights of 5.2 kg, 3.5 kg, 7 kg, and 9.74 kg.

(a) What is the total weight of the dogs?

(b) What is the average weight of the four dogs?

5. The table shows how far, in miles, Darryl ran the last 5 days. What was the average number of miles Darryl ran each day? Round your answer to one decimal place.

Monday	Tuesday	Wednesday	Thursday	Friday
7.2 m	5.5 m	4 m	8.7 m	5 m

6. On a 5-day vacation, the Salvador family drove an average of 53 miles each day. How many miles in total did they drive?

7. The large dogs in a kennel eat a total of 432 ounces of dry food every day. The average amount one dog eats is 24 ounces per day. How many large dogs are in the kennel? Show your work.

8. The mean of 5 numbers is 12.4. Four of the numbers are 18.8, 12.6, 8.25, and 14. What is the fifth number?

9. Each week, Alexa's company deducts money from her paycheck and places it into a savings account. She saves $75.50 each week. How many weeks will it take to save $528.50?

10. The average length of five bean seedlings is 7.3 cm. Refer to the table and determine the missing seedling's length. Show your work.

Length in cm	5.3	8.2	9.1	6.4	?

11. The Clark family spends $287.40, on average, each month for gas. Use the table to find out how much they spent in May. Show your work.

Month	Jan.	Feb.	March	April	May	June
Amount	$302.15	$243.65	$263.55	$305.45	?	$332.90

12. Jack has $425 less in savings than Mattias. If the average of the two men's savings is $2,100, how much money does Mattias have in savings?

13. The average height of three trees is 16.4 ft. The first tree is $\frac{1}{2}$ the height of the second tree. The sum of the heights of the first two trees is $\frac{1}{2}$ of the total height of all three trees. Find the height of the second tree.

14. Arman and Brianna have a total of 87 stickers. Brianna and Carlos have a total of 111 stickers. Carlos has five times as many stickers as Arman. What is the average number of stickers that Arman and Carlos have?

Basics

1. Calista ran 6.75 miles in 45 minutes. At this rate, how far can she run per minute?

2. In an hour, a machine can pack 450 cans of tennis balls. At this rate, how many cans are packed in one minute?

3. A one-bedroom apartment in a city in Washington costs $1,650 per month. At that rate, how much does it cost to live in the apartment each day in the month of June?

4. The Ivanov family budgets $15,420 each year for rent. At this rate, how much do they pay for three months' rent?

5. A machine can fill 252 cream puffs in 45 minutes. How long will it take to fill 448 cream puffs?

6. 5 m of wire costs \$12.35. How much does 2 m of wire cost?

7. Two typists type the same 6,000 word essay. The first typist types 40 words per minute. The other typist types 75 words per minute. How many fewer minutes does it take the faster typist to type the essay?

8. An appliance repairman charges a service fee of $25, and then $48.50 per hour. For every fraction of an hour less than one hour, he charges that fraction of $48.50. If it took the repairman 2 and a half hours to fix a dishwasher, how much money did he charge? Show your work.

9. A delivery service charges $8.00 for pick-up, and then $3.85 per mile. A company wants to send a package 17 miles across town. How much will it cost?

10. Andrei makes $18.80 per hour working at a furniture store. After 40 hours in one week, he makes 1.5 times his regular wage. How much did Andrei make working 52 hours in one week? Show your work.

11. A race car can travel 81 km in 20 minutes.

(a) At this rate, how far can it travel in 45 minutes?

(b) At this rate, how long would it take to travel 486 km?

12. 2 liters of high quality olive oil, 4 liters of medium quality olive oil, and 4 liters of low quality olive oil were combined to form everyday olive oil. The costs of high, medium, and low quality olive oil, per liter, were $12.50, $8.75, and $7.50, respectively. Find the cost of 3 liters of everyday olive oil.

13. A printer can print 300 pages in 15 min. How many hours will it take three similar printers to print 4,800 pages? Express your answer in hours and minutes.

6 Rate

6.3 Speed

Basics

1. On a 5-hour bike tour, a cyclist bikes 56 miles. What is the average speed of the cyclist on the trip?

2. The Grants drove 243 miles in 4.5 hours. What was their average speed?

3. Mrs. Jung is taking a $6\frac{1}{2}$ mile walk. She walks at an average speed of 2.6 miles per hour. How long will it take her to finish her walk?

4. A bus traveled at an average speed of 49 miles per hour for 3 hours and 45 minutes. How far did the bus travel?

5. It takes Ashimah 45 minutes to bike from home to school. Her average speed is 14 miles per hour. How many miles is it from home to school?

6. A bus traveled from City A to City B, and then from City B to City C. The distance from City A to City B is 225 miles. The bus traveled this distance at an average speed of 50 miles per hour. The distance from City B to City C was 135 miles, which the bus traveled at an average speed of 54 miles per hour. How long did it take for the bus to travel from City A to City C?

7. Alyssa drove her car for $2\frac{1}{4}$ h and traveled 217.35 km. If she drives her car at the same speed for 3 h, how many km will Alyssa travel?

8. A truck left Town A at 9:00 am, driving at an average speed of 70 mph. At 11:15 am, the truck reached Town B.

(a) What is the distance between Town A and Town B?

(b) How long would it take the truck to drive from Town B to Town A at an average speed of 50 mph?

9. Tony and Sumin both left Town A at 5:00 p.m., heading for Town B. Tony's average speed was 15 mph faster than Sumin's average speed. Tony reached Town B at 8:30 p.m. How many miles was Sumin from Town B at 8:30 p.m.?

10. At 9:00 a.m., Ella left Town P towards Town R at a speed of 60 mph. At 10:00 a.m., Pablo left Town R towards Town P and his speed was 10 mph slower than Ella. At 10:30 a.m., Ella had traveled $\frac{3}{5}$ of the journey. At what time did Pablo arrive at Town P?

7 Percent

7.1 Meaning of Percent

Basics

1. Look at the 100-grid and express the shaded units as a fraction, decimal, and percent. Express your answers in simplest form.

(a)

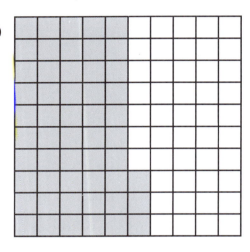

Fraction   _____

Decimal   _____

Percent   _____

(b)

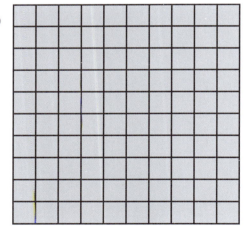

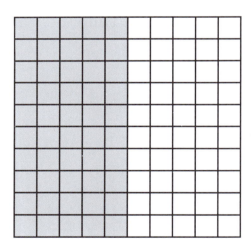

Fraction   _____

Decimal   _____

Percent   _____

2. Shade the following on the grid.

(a) $\frac{47}{100}$ of the units

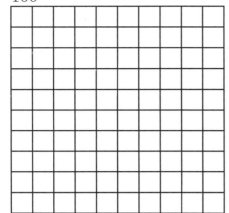

(b) 0.68 of the units

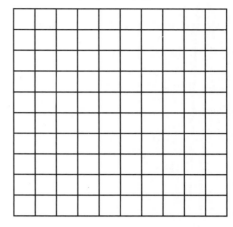

(c) $\frac{7}{5}$ of the units

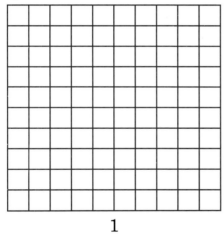

1

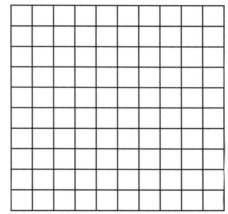

(d) 99%

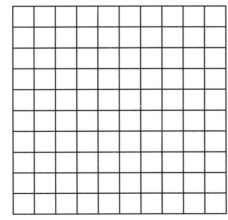

3. $\frac{3}{4}$ of the vehicles in a parking lot are trucks. What percentage of the vehicles are trucks?

4. In a dog park, 11 out of 20 of the dogs are beagles. What percentage of the dogs are beagles?

5.   Ms. Sinclair donated $\frac{3}{8}$ of her work bonus to charity and kept the rest. What percentage of Ms. Sinclair's work bonus did she keep?

6.   Raj intended to run a certain distance on Saturday. He ran the entire distance, and then half of the way back to the starting point. What percentage of the distance that Raj intended to run did he actually run on Saturday?

7.   Express each fraction as a percent. For any repeating decimals, round to one decimal place.

(a) $\frac{2}{3}$ _____

(b) $\frac{4}{9}$ _____

8. Complete the table.

	Fraction in Simplest Form	Decimal	Percent
(a)	$\frac{1}{2}$	0.5	50%
(b)	$\frac{3}{4}$	0.75	
(c)	$\frac{3}{10}$		
(d)		0.58	
(e)			$3\frac{1}{2}\%$
(f)			225%
(g)			505%
(h)	$\frac{428}{100}$		
(i)		0.44	

9. 60% of the people at a concert are children. What percentage of the people at the concert are not children?

7 Percent

7.2 Percentage of a Quantity

Basics

1. Find the value of each.

   (a) 60% of $90   _____

   (b) 75% of $90   _____

   (c) 5% of $90   _____

   (d) 25% of 180 inches   _____

   (e) $33\frac{1}{3}$% of 105 km   _____

   (f) 75% of 184 cm   _____

   (g) $8\frac{1}{2}$% of $200   _____

2. **(a)** What percent is 13 out of 25?

    (b) What percent is 105 out of 300?

3. Aurora saved $850. She used 35% of her savings on a new television. How much did the television cost?

4. In a pet shop, 80% of the pets are mammals. If there are 65 pets in the store, how many are not mammals?

5. Express 50 cm as a percentage of 10 m.

6. Amanda found 2 blouses on sale. The blue one has a regular price of $97, and is marked 25% off. The red blouse is 20% off the original cost of $88. Which blouse costs less and by how much?

7. The Nelson Family owns an orchard. 43% of the trees in the orchard are apple trees. If there are 387 apple trees, how many trees are in the orchard?

8. Heather took a day-long hike. She hiked 38% of her hike before lunch. If she hiked 4.56 miles before lunch, how many miles did she hike after lunch?

9. A video game store had 192 games left after selling 68% of its game inventory. How many games did the store have at first?

10. At a car dealership, there are vans, trucks, and sedans on the lot. 16% of the inventory are trucks and 24% of the inventory are vans. There are 24 trucks and 36 vans. The rest are sedans. How many vehicles are on the lot?

11. A pair of pants is discounted 25%. They now cost $104.25. What was the cost before the discount?

12. Regular admission price for a family to enter an amusement park is $345.80. The Patel family used a discount coupon that gave them 35% off the normal price. How much did the family save?

13. Last year, a school had a student population of 1,960 students. This year, the number of students at the school increased 20%. How many students are now attending the school?

14. A new car cost $16,662.50 including 7.5% tax. What was the price before tax?

15. One day at a dairy bar, 25% more ice cream was sold than frozen yogurt. If 375 servings of ice cream were sold, how many total servings of frozen yogurt and ice cream were sold?

16. If 25% of the first number is 125 and 35% of the second number is 10.5, what is the sum of both numbers?

17. Ms. Seville always donates 10% of her annual income to charity. Last year, her income increased 20% when compared to the prior year, and Ms. Seville gave $768 more to charity than the year before. Find Ms. Seville's income last year.

18. There are 20 children on a school bus. 60% of the children were boys. After some more girls got on the bus, 40% of the children were boys. How many more girls got on the bus?

Answer Key

1.1 Order of Operations

1. (a) expression   (b) equation
   (c) expression   (d) expression
   (e) equation

2. (a) sum   (b) difference
   (c) product   (d) quotient

3. (a) 3   (b) 2
   (c) 1

4. (a) $3 \times 12 - 5 = 31$
   (b) $8 \times 1 + 2 = 10$
   (c) Answers may vary; 19

5. Answers will vary.

6. (a) Answers will vary.
   (b) Answers will vary.

7. Base, exponent

8. 4
   2, 2, 2
   4, 2, 2, 2, 2
   2^5, 2, 2, 2, 2, 2, 32

9. 1, 1, 1, 1, 1
   10, 10, 10, 1,000
   100, 100, 10,000

10. (a) $9^2 = 81$ (b) $8^3 = 512$
    (c) $3^5 = 243$ (d) $3^3 \times 4^2 = 432$
    (e) $5^2 \times 2^4 = 400$

11. (a) 64   (b) 128
    (c) 400   (d) 216

12. Leo thought 6^3 meant 3 sixes. The sum of 3 sixes, that is, 3 groups of six, is 18. But 6^3 means the product of 6 multiplied by itself 3 times: $6 \times 6 \times 6 = 216$.

13. 9

14.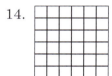

15. (a) 2   (b) 5
    (c) 3

16. (a) is the correct equation.
    The Order of Operations says we multiply before we add. The second equation was solved without considering the Order of Operations.

17. (a) 14   (b) 15
    (c) 1

18. (a) $14 + 12 \times 36$ or $12 \times 36 + 14$
    (b) $12 + 14 \times 36$ or $14 \times 36 + 12$
    (c) 70 more prizes

19. (a) Third   (b) First
    (c) Second

20. (a) 10   (b) 528
    (c) 69   (d) 176
    (e) 7

21. (a) $7 \times 12 + 3 \times 12 - 4 - 2 = 114$

    (b) Answers will vary.

22. (a) 12   (b) 7

    (c) 20   (d) 2

23. (a) $48 \div (3 + 5)$

    (b) $(5 \times 9) \div (5 \times 3)$

    (c) $2 \times 3 \times (15 - 9)$

24. $2^2 + 2 \times (3 + 3^2) \div 3$ matches to 12

    $(2^2 + 2) \times (3 + 3^2) \div 3$ matches to 24

    $2^2 + (2 \times 3) + 3^2 \div 3$ matches to 13

25. (a) 24   (b) 40

    (c) 17

26. (a) $6 \times (3 + 3^3) = 180$

    (b) $1 = (1 + 39) \div (16 - 6) \div 2^2$

    (c) $(5 + 7) \times (8 - 6) \times 2 = 48$

27. (a) Answers for showing work may
        vary; $5

    (b) $7 \times 15 + 11 = 116$ chairs

28. Cora is correct; Alyssa solved this by
    reading from left to right, without
    considering the Order of Operations.

29. 15, 21

30. 5,050

1.2 Factors and Multiples

1. (a) 1, 2, 7, 14

   (b) 1, 2, 4, 8, 16, 32

   (c) 1, 2, 3, 4, 6, 8, 12, 16, 24, 32, 48, 96

   (d) 1, 3, 5, 7, 15, 21, 35, 105

2. (a) 9   (b) 14

   (c) 18

3. (a) 1, 2   (b) 1, 7

4. 7 arrangements; each arrangement will
   have 3 yellow flowers, 5 pink flowers,
   and 6 white flowers.

5. 12 m long

6. 4 bags

7. Answers may vary.

8. 12

9. (a) 7, 14, 21, 28, 35

   (b) 12, 24, 36, 48, 60

   (c) 24, 48, 72, 96, 120

10. 3 {3, 6, 9, ⑫ 15, 18, 21, 24, 27, 30}
    4 {4, 8, ⑫ 16, 10, 24, 28, 32, 36, 40}

11. (a) 36   (b) 180

12. 60

13. Answers will vary.

14. 24 days

15. 24th caller

16. 84 in

17. 9:12 a.m.

18. 60 minutes

19. 167

1.3 Multiplication

1. (a) Distributive Property of
        Multiplication

    (b) Associative Property of
        Multiplication

    (c) Identity Property of
        Multiplication

    (d) Commutative Property of
        Multiplication

    (e) Zero Property of Multiplication

2. (a) 560   (b) 360

    (c) 630   (d) 168,000

3. (a) 7, 7   (b) 9, 9

    (c) The Distributive Property of
        Multiplication

4. 95 m²

5. (a) $40 \times \$2 + 3 \times (10 \times \$2) + 5 \times (2 - \$0.50)$

    (b) Jenna does not have enough money.

6. (a) Solution methods will vary; 144

    (b) Solution methods will vary; 520

7. (a) 476   (b) 232

8. (a) 126,000

9. Solution methods will vary.

(a) 228 (b)   252

    (c) 1,008   (d) 343

10. Solution methods may vary.

    (a) 48,000   (b) 581

11. Answers may vary.

    (a) $16 \times 7 = (8 \times 7) + (8 \times 7)$

        $= 56 + 56 = 112$

    (b) $16 \times 7 = (10 \times 7) + (6 \times 7) =$

        $70 + 42 = 112$

    (c) $16 \times 7 = (20 \times 7) - (4 \times 7) =$

        $140 - 28 = 112$

    (d) $16 \times 7 = (16 \times 10) - (16 \times 3) =$

        $160 - 48 = 112$

    (e) $16 \times 7 = (16 \times 2) + (16 \times 2)$

        $+ (16 \times 2) + 16 = 32 + 32 + 32 + 16$

        $= 112$

    (f) $16 \times 7 = (16 \times 5) + (16 \times 2) = 80 + 32$

        $= 112$

1.4 Division

1. Dividend, divisor, quotient

2. (a) 6 yards

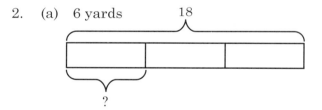

    (b) 6 curtains

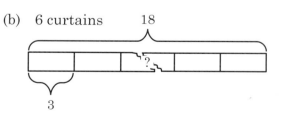

3. The quotient does not change.
    Examples will vary.

4. (a) 7, 3   (b) 8, 6

    (c) 2, 5   (d) 4, 4

5.   7 bags

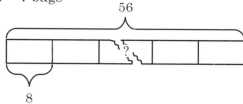

6.   8 in

7.   When we divide a number by 1, we get the same number. $100 \div 1 = 100$

When we divide a number by itself, we get 1. $100 \div 100 = 1$

When we divide 0 by a non-zero number, we will get 0. $0 \div 100 = 0$

8.   (a)   15, 15, 15   (b)   50, 1

9.   (a)   12   (b)   34
     (c)   72   (d)   57
     (e)   300

10. Explanations may vary.
     (a)   True.   (b)   True.

11. Answers may vary.

12. (a)   16   (b)   68
     (c)   15   (d)   70
     (e)   11,000

13. Both girls' methods work because Riya ended up dividing by 6, just like Mariam.

14. Tim is correct. Explanations may vary.

15. (a)   63   (b)   90
     (c)   9,000

2.1 Multiplication of Fractions

1. $\frac{3}{5}$

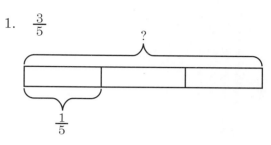

2. $1\frac{1}{2}$

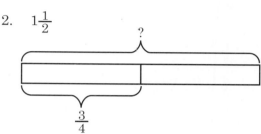

3. $2\frac{1}{4}$ c

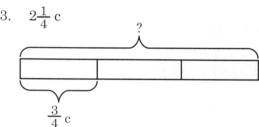

4. $2\frac{2}{5}$ m

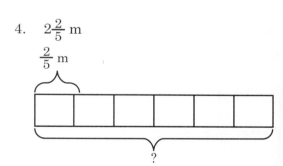

5.   32 miles

6.   115 female friends

7.   87 of the cars are used.

8.   45 boys

9. $212

10. (a) $\frac{1}{12}$ (b) $\frac{2}{15}$
     (c) $\frac{1}{2}$ (d) $\frac{4}{7}$

11. $\frac{3}{10}$

12. $\frac{1}{49}$

13. $\frac{2}{3}$ m is painted red.

$\frac{1}{12}$ m of the rod is not painted.

14. $\frac{23}{24}$

15. (a) $4\frac{1}{5}$ (b) $15\frac{1}{2}$

    (c) $8\frac{1}{2}$ (d) $31\frac{1}{2}$

16. $10\frac{2}{3}$ kg

17. $28\frac{1}{2}$ oz

18. $7\frac{1}{2}$ lb

19. 210 cards

20. 700 history books

21. (a) $3\frac{5}{8}$ (b) $1\frac{13}{32}$

    (c) $\frac{17}{18}$ (d) $1\frac{1}{3}$

22. (a) 14   (b) $2\frac{6}{7}$

    (c) $14\frac{3}{5}$

23. $2\frac{3}{4}$ tons

24. $17\frac{2}{3}$ ft^2

25. $\frac{2}{5}$ L

2.2 Division of Fractions

1. (a) 24   (b) 8

    (c) $4\frac{4}{5}$

2. (a) 9   (b) $31\frac{1}{2}$

    (c) 25

3. 8 stops

4. 45 bags

5. $8\frac{1}{3}$ c

6. 5 small containers

7. $\frac{2}{3} \div 5$

8. (a) $\frac{1}{25}$ (b) $\frac{5}{12}$

    (c) $\frac{1}{25}$

9. $\frac{1}{6}$ m

10. $\frac{1}{6}$ of the pan

11. $\frac{1}{8}$ lb

12. $\frac{7}{81}$ yd

13. $\frac{2}{15}$ of her savings

14. (a) 3   (b) 2

    (c) 3   (d) 1

15. (a) 7   (b) $3\frac{1}{2}$

    (c) $1\frac{2}{5}$ (d) $1\frac{1}{6}$

16. (a) 2   (b) 2

    (c) $4\frac{1}{2}$ (d) $2\frac{3}{4}$

17. (a) $\frac{6}{7}$ (b) $\frac{15}{28}$

    (c) $4\frac{1}{2}$ (d) $1\frac{1}{24}$

18. Quotient = $1\frac{1}{2}$

Example problem will vary.

19. 7 pieces

20. $\frac{9}{10}$ of a mile

21. $4\frac{6}{11}$ ft

22. $29\frac{6}{25}$ ft

3.1 Addition and Subtraction of Decimals

1. (a) 3, 5   (b) 0, 0, 1
   (c) 6, 0, 2   (d) 8, 7
   (e) 0, 4, 4

2. (a) 0.9   (b) 2.3
   (c) 0.45   (d) 5.2
   (e) 3.05   (f) 0.032

3. (a) 0.75   (b) 0.88
   (c) 3.2   (d) 1.125
   (e) 3.75   (f) 4.75

4. (a) $6\frac{17}{100}$ (b) $3\frac{3}{5}$
   (c) $\frac{6}{25}$ (d) $13\frac{9}{40}$
   (e) $2\frac{9}{20}$ (f) $8\frac{7}{250}$

5. (a) 0.302   (b) 4.5
   (c) 20.372

6. (a) $0.001, \frac{1}{100}, 0.1, 1\frac{1}{10}$
   (b) $16.02, 16.022, 16\frac{5}{25}, 16\frac{22}{100}$

7. Answers will vary.

8. (a) 21.305   (b) 77.771
   (c) 27.025   (d) 5.399

9. (a) 1.56   (b) 2.25
   (c) 1.234   (d) 0.225

10. $0.26

11. No. She needs 2.171 meters more.

12. 14.98 miles

13. (a) 3; 0; 7; 4   (b) 3,074
    (c) 307; 4   (d) 30; 7; 4

14. 11.96, 16.22

15. 1.58 km

16. $1,006

17. $2.75

3.2 Multiplication of Decimals

1. (a) 0.05   (b) 0.5
   (c) 5   (d) 50
   (e) 5   (f) 0.5

2. (a) 400   (b) 4,200
   (c) 7.17   (d) 7.17
   (e) 900   (f) 0.009

3. (a) $485   (b) $43.65

4. $8,713.50

5. (a) Less than 205.63
   (b) Greater than 205.63
   (c) Less than 205.63
   (d) Greater than 205.63

6. (a) =   (b) >
   (c) <   (d) <
   (e) >   (f) <
   (g) =   (h) >
   (i) >

7. (a) 4.5   (b) 56.1
   (c) 0.35   (d) 0.0782
   (e) 13.12

8. (a) 1,806.25 square inches
   (b) 19.683 cubic meters
   (c) $58.46

9. 42.84 cm shorter

10. $1,980

11. $400.85

3.3 Division of Decimals

1. (a) Less than the dividend
    (b) Greater than the dividend
    (c) Less than the dividend
    (d) Less than the dividend

2. (a) =   (b) <
    (c) <   (d) <
    (e) <   (f) >
    (g) >   (h) >
    (i) >

3. (a) 3   (b) 7.2, 10
    (c) 6   (d) 3,090

4. (a) 8   (b) 0.8
    (c) 0.08   (d) 0.8
    (e) 800

5. (a) 0.5   (b) 1.5
    (c) 12.4   (d) 68.2
    (e) 32.5   (f) 2.6
    (g) 3.1

6. 8 tons

7. 21.2 meters

8. 21 bags

9. The jumbo size

10. $165

3.4 Metric Measurements and Decimals

1. (a) 15,000   (b) 1,500
    (c) 150   (d) 15,000
    (e) 7,000   (f) 700
    (g) 62,000   (h) 6,200

2. (a) 0.005   (b) 0.0052
    (c) 0.013   (d) 0.0138
    (e) 0.0008   (f) 0.0088
    (g) 10.602   (h) 1.2

3. 3 cushions

4. (a) 16.83 kg   (b) 7,520 g

5. 575 mL

6. 2,800 grams

7. 1.15 L

4.1 Positive and Negative Numbers

1. (a) sea level, −3,000 m
    (b) 0°F, +85°F

2. 20 m above sea level

3. (a) −4   (b) 2

4. Bird A: 0.05 g   Bird B: 0.02 g
    Bird C: −0.01 g   Bird D: −0.005 g
    Bird E: 0.001 g

5. Ada: −$6.87   Briana: $2.08
    Cooper: −$18.85   Darryl: $15.37
    Ethan: $8.27

6. 49 Palms Oasis: −2.34 miles

Split Rock Loop: −2.84 miles

Lost Palms Oasis: 2.16 miles

Lost Horse Loop: 1.16 miles

Willow Hole: 1.86 miles

7. (a) The Canton Tower

157 meters taller

(b) Relative position: −157 m

8. (a) Relative position: −40 feet

(b) Relative position: −40 feet

(c) The distance between the
Roosevelt Island or the Lexington
Avenue stations and the 191st
Street station offers the greatest
distance. Any of these three could
be the reference point.

4.2 Comparing Positive and Negative Numbers

1. (a)

(b) −5, −2$\frac{1}{2}$, 0, 2.5

(c) Answers will vary.

2. (a)

$$
\begin{array}{c}
2 \\
1\frac{1}{2} \mathbin{\rlap{\hspace{2pt}}} 1 \mathbin{\rlap{\hspace{2pt}}} 1\frac{1}{5} \\
0 \\
-1 \\
-2 \\
-2.4 \\
-3 \mathbin{\rlap{\hspace{2pt}}} -3.2 \\
-4
\end{array}
$$

(b) −3.2, −2.4, 1$\frac{1}{5}$, 1$\frac{1}{2}$

3. (a) >   (b) <

(c) >

4. (a) 2,675 m   (b) −10 °F

(c) −$1,503

5. Ani: 6, Shanice: −2, Nolan: 5, Tyler: −8,
Mia: −10

6. 2$\frac{1}{12}$, $\frac{5}{6}$, $\frac{1}{2}$, 0, $\frac{1}{3}$, −1, −2$\frac{1}{2}$, −2$\frac{7}{24}$, −1$\frac{2}{3}$

7. 6, 4$\frac{1}{2}$, 0.4, −0.4, −4.25, −4$\frac{1}{3}$

8. Answers will vary.

9. 443 m

10. (a) 4   (b) 4

(c) 6   (d) 6

(e) 8

11. (a) <   (b) >

(c) <   (d) >

(e) =   (f) >

12. (a) >   (b) <

(c) >   (d) <

(e) =   (f) <

13. |−0.8|, |1|, |7.2|, |−72.3|, |73.2|

14. |21|, |12|, |−12|, |−11|, |8|, |−8|

15. (a) 50 points   (b) −20 points

(c) 8 correct, 2 unanswered

(d) 8 correct, 2 incorrect

16. (a) −4   (b) 8

5.1 Ratios and Equivalent Ratios

1. (a) 7 : 2   (b) 2 : 1
   (c) 2 : 7   (d) 1 : 2 : 7
   (e) 7 : 10
   (f) Answers will vary.

2. (a) 3 : 8   (b) 8 : 3
   (c) 3 : 11

3. (a) 3 : 2   (b) 2 : 3
   (c) 5 : 2

4. (a) 7 : 8   (b) 8 : 15

5. 9 : 2

6. 37 : 60

7. (a) 13 : 30   (b) 3 : 43

8. $60.00

9. 15 : 11

10. (a) 4 : 3   (b) 3 : 4

11. (a) 8 : 4   (b) 2 : 3
    (c) 35 : 25   (d) 100 : 30

12. (a) 1 : 3   (b) 2 : 3
    (c) 1 : 4   (d) 4 : 7 : 1
    (e) 4 : 3   (f) 8 : 9
    (g) 7 : 5   (h) 2 : 6 : 5

13. (a) 9   (b) 8
    (c) 8   (d) 10, 5

14. (a) 2 : 5   (b) 5 : 7

15. 5 : 2 : 7

16. 7 : 9 : 5

17. (a) 78 : 27 or 26 : 9
    (b) 27 : 78 or 9 : 26
    (c) 26 : 9   (d) No

18. 48 cm

19. 54 m

20. $64

21. 30 girls

5.2 Ratios and Fractions

1. (a) $\frac{1}{4}$ (b) $\frac{2}{3}$
   (c) $\frac{14}{5}$ (d) $\frac{5}{2}$

2. $\frac{5}{4}$ or $1\frac{1}{4}$

3. 3 : 5

4. (a) $\frac{7}{16}$ (b) $\frac{9}{16}$
   (c) 21

5. (a) $\frac{3}{8}$ (b) $\frac{5}{8}$
   (c) 25 spoons

6. Paula: $144, Kawai: $168

7. 12 : 2 : 1 : 16

8. (a) 5 : 9   (b) 100 cm

9. 6 more baseball cards.

10. (a) $\frac{3}{6} = \frac{1}{2}$ (b) 21 years older

11. 135 road bikes

12. $\frac{5}{11}$

6.1 Average and Rate

1. Yes

2. 47.2

3. (a) 72 cm   (b) 4.59 L

4. (a) 25.44 kg   (b) 6.36 kg

5. 6.1 miles per day

6. 265 miles

7. 18 large dogs

8. 8.35

9. 7 weeks

10. 7.5 cm

11. $276.70

12. $2,312.50

13. 16.4 ft

14. 18

6.2 Unit Rate

1. 0.15 miles per minute

2. 7.5 cans

3. $55

4. $3,855

5. 80 minutes

6. $4.94

7. 70 fewer minutes

8. $146.25

9. $73.45

10. $1,090.40

11. (a) 182.25 km

     (b) 120 minutes or 2 hours

12. $27

13. 1 hr 20 min

6.3 Speed

1. 11.2 miles per hour

2. 54 miles per hour

3. 2.5 hours

4. 183.75 miles

5. $10\frac{1}{2}$ miles

6. 7 hours

7. 289.8 km

8. (a) 157.5 mi   (b) 3.15 h

9. 52.5 miles

10. 1:00 p.m.

7.1 Meaning of Percent

1. (a) $\frac{53}{100}$, 0.53, 53%

     (b) $1\frac{1}{2}$, 1.5, 150%

2. (a)

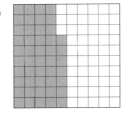

(b)

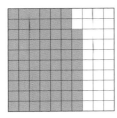

(c)

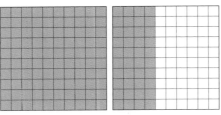

(d)

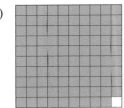

3. 75%

4. 55%

5. 62.5%

6. $1\frac{1}{2}$ times or 150%

7. (a) 66.6%   (b) 44.4%

8. (a) Already completed

(b) 75%   (c) 0.3; 30%

(d) $\frac{29}{50}$; 58%

(e) $\frac{7}{200}$; 0.035

(f) $2\frac{1}{4}$; 2.25

(g) $5\frac{1}{20}$; 5.05

(h) 4.28; 428%   (i) $\frac{33}{75}$; 44%

9. 40%

7.2 Percentage of a Quantity

1. (a) $54.00   (b) $67.50

(c) $4.50   (d) 45 inches

(e) 35 km   (f) 138 cm

(g) $17.00

2. (a) 52%   (b) 35%

3. $297.50

4. 13 pets

5. 5%

6. The red blouse costs $2.35 less.

7. 900 trees

8. 7.44 miles

9. 600 games

10. 150 vehicles

11. $139.00

12. $121.03

13. 2,352

14. $15,500.00

15. 675

16. 530

17. $46,080

18. 10